Pattern Parade

Copyright © Gareth Stevens, Inc. All rights reserved.

Developed for Harcourt, Inc., by Gareth Stevens, Inc. This edition published by Harcourt, Inc., by agreement with Gareth Stevens, Inc. No part of this publication may be reproduced or transmitted in any form or by any means, electronic or mechanical, including photocopy, recording, or any information storage and retrieval system, without permission in writing from the copyright holder.

Requests for permission to make copies of any part of the work should be addressed to Permissions Department, Gareth Stevens, Inc., 330 West Olive Street, Suite 100, Milwaukee, Wisconsin 53212. Fax: 414-332-3567.

HARCOURT and the Harcourt Logo are trademarks of Harcourt, Inc., registered in the United States of America and/or other jurisdictions.

Printed in the United States of America

ISBN 13: 978-0-15-360168-2
ISBN 10: 0-15-360168-X

1 2 3 4 5 6 7 8 9 10 039 16 15 14 13 12 11 10 09 08 07

Pattern Parade

by Joan Freese

Photographs by Kay McKinley

SCHOOL PUBLISHERS

It is Pattern Week in Miss Penny's class. The class will learn about different patterns. The children ask Miss Penny, "What is a pattern?"

"Let me show you," says Miss Penny.

triangle, square, circle, triangle, square, circle, triangle, square, circle

Miss Penny draws a blue triangle on the board. Then she adds a blue square. Next she adds a blue circle. She repeats the three shapes to make a pattern.

purple, blue, purple, blue, purple, blue

Next, Miss Penny lines up balls on the table. Purple, blue, purple, blue. She asks the class what color comes next. "A purple ball," Toby answers. The colors make a pattern.

blue, yellow, blue, yellow, blue, yellow

The class likes the patterns Miss Penny shows them. They want to make their own. Jack makes a pattern from blocks he finds in the classroom. "Good job!" says Miss Penny.

blue, white, blue, white, blue, white

Miss Penny asks the class to look for other patterns. Simone says the floor tiles have a pattern. "What is the pattern?" Miss Penny asks. "Blue, white, blue, white. The colors repeat," says Simone.

big, small, big, small, big, small

Now it is time for recess. The class lines up near the wall. The children will play outside. They are holding big and small balls. The balls make a pattern.

1, 2, 3, 1, 2, 3, 1, 2, 3

Later Miss Penny shows the class another pattern. This pattern uses numbers. Miss Penny writes the number pattern on the board. There are three numbers in this pattern unit.

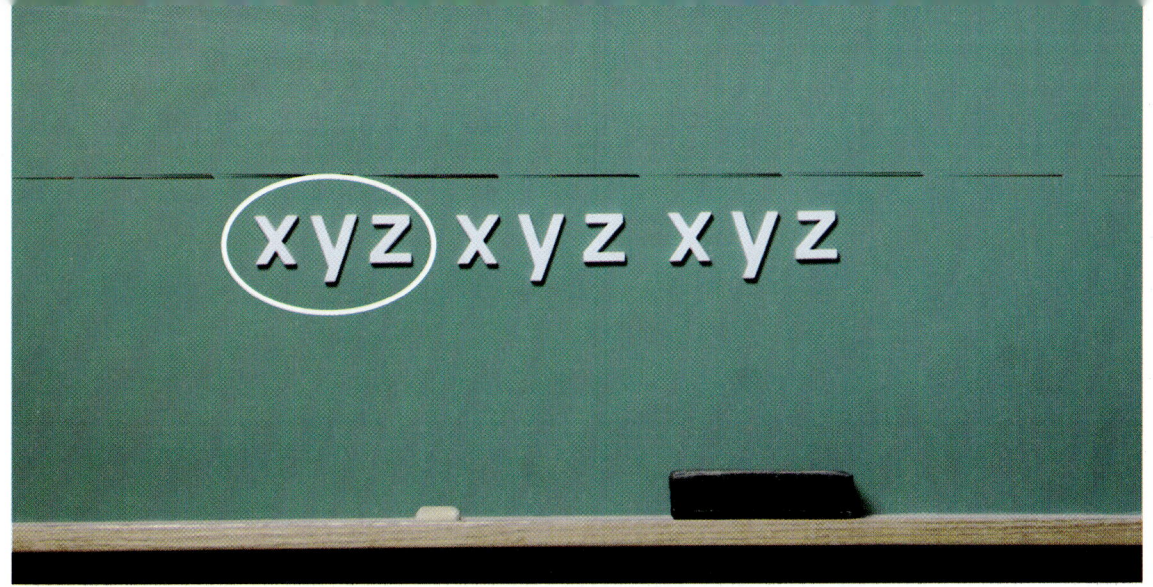

x, y, z, x, y, z, x, y, z

Miss Penny asks the class to make a pattern using letters. Henry creates a pattern from the letters x, y, and z. Miss Penny says, "Nice work! Your pattern unit is three letters long."

green, blue, yellow, green, blue, yellow, green, blue, yellow

The class looks for patterns in other places. Mimi spots Jamie's striped sweater. She says, "Jamie's top has stripes." The colors make a pattern.

coat, hat, bag, coat, hat, bag, coat, hat, bag

Mark looks around the classroom for a pattern. He does not have to look far! He sees hats, coats, and bags hanging in the closet. The things make a pattern.

apple, apple, orange, apple, apple, orange, apple, apple, orange

The class is busy searching for patterns. Miss Penny shows the class some fruit. They see a pattern right away. The children tell Miss Penny what comes next.

white, white, black, white, white, black, white, white, black

Now it's time for gym class. The children line up their right shoes in a row. The colors of the shoes make a pattern. "Patterns are everywhere!" Juan says.

13

short, short, tall, short, short, tall, short, short, tall

Next, the class looks around for more patterns. Iris sees boxes on a shelf. Some boxes are tall and other boxes are short. The boxes make a pattern.

pink, purple, purple, pink, purple, purple, pink, purple, purple

Lisa sees crayons in the art room. She has an idea. She forms a pattern with the crayons. Her friends help her pick what color comes next.

stand, stand, sit, stand, stand, sit, stand, stand, sit

Miss Penny has another pattern to share. "This pattern is three children long. Will you help me make it?" she asks. "Yes," says the class. Everyone wants to help.

front, front, back, front, front, back, front, front, back

Then the class makes its own pattern. Everyone stands in a row. Two children face the front. One child faces the back. Then they repeat the pattern.

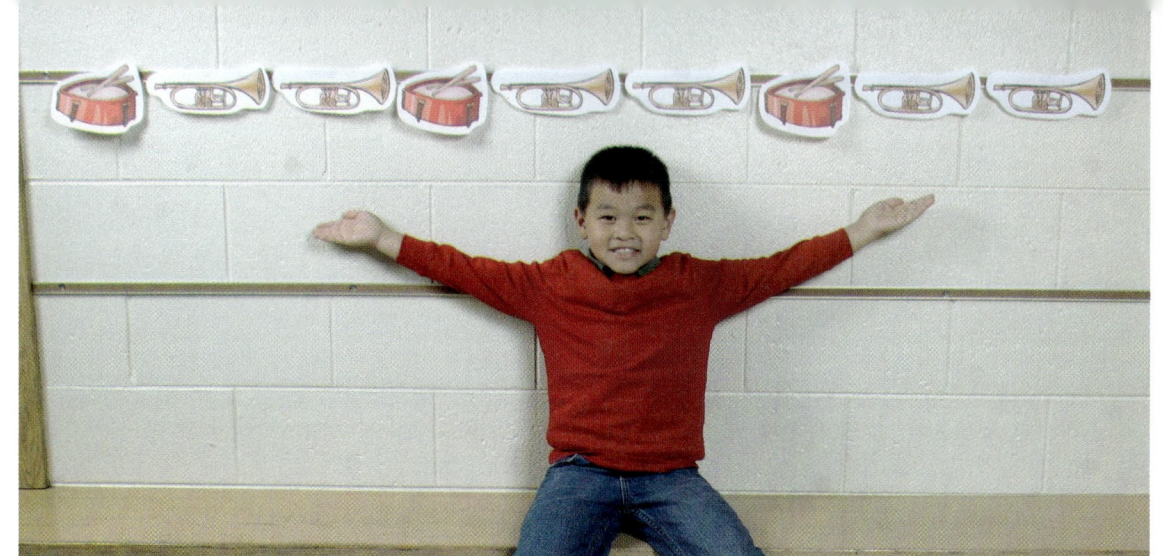

drum, horn, horn, drum, horn, horn, drum, horn, horn

    Miss Penny's class is getting very good with patterns. Jake spots another pattern. It is in the music room. He says, "The pictures of the drums and horns make a pattern."

white, blue, blue, white, blue, blue, white, blue, blue

Mimi doesn't have to go far to find a pattern. "Look at my necklace," she says to the class. The beads in Mimi's necklace are blue and white. The colors make a pattern.

It is sharing day in Miss Penny's class. The children bring things with patterns to school. The class has a pattern Show and Tell. You can see patterns in clothes and games.

At the end of the day the class lines up to go home. Miss Penny gives each child a crown to wear. The class helps make a pattern. It is Miss Penny's pattern parade!

Glossary

parade to march

repeat to do or say something again

tile a thin, hard plate used to cover a floor or roof